YOUR KNOWLEDGE HAS VALUE

Andrew Magdy Kamal

Quantum Field Theory, A Theoretical Framework

GRIN Verlag

Bibliografische Information der Deutschen Nationalbibliothek:

Die Deutsche Bibliothek verzeichnet diese Publikation in der Deutschen National-
bibliografie; detaillierte bibliografische Daten sind im Internet über http://dnb.d-
nb.de/ abrufbar.

Imprint:

Copyright © 2013 GRIN Verlag GmbH
Druck und Bindung: Books on Demand GmbH, Norderstedt Germany
ISBN: 978-3-656-46711-3

This book at GRIN:

http://www.grin.com/en/e-book/230367/quantum-field-theory-a-theoretical-frame-
work

GRIN - Your knowledge has value

Der GRIN Verlag publiziert seit 1998 wissenschaftliche Arbeiten von Studenten, Hochschullehrern und anderen Akademikern als eBook und gedrucktes Buch. Die Verlagswebsite www.grin.com ist die ideale Plattform zur Veröffentlichung von Hausarbeiten, Abschlussarbeiten, wissenschaftlichen Aufsätzen, Dissertationen und Fachbüchern.

Visit us on the internet:

http://www.grin.com/

http://www.facebook.com/grincom

http://www.twitter.com/grin_com

Andrew
Nassif
2013

Quantum Field Theory

A Theoretical Framework

A research paper on some of the biggest controversies in the field of science and theoretical physics. This includes attempts on finding new scientific discoveries and research on the field itself as well as what is Quantum Field Theory.

Quantum Field Theory is the framework of all Quantum models in Quantum Physics and Theoretical Physics. It goes on the bases of having many bodies in the form of condensed matter. This helps us realize what we are made of, how the universe works, and what builds us up. These fundamentals are important in understanding the world around us, which is what science is meant to do. Particle Physics itself is revised by theories in Quantum Field Theory. This also sheds light on the area of both Quantum Mechanics and Elementary Physics. In the field Quanta there are ripples of matter and specs of atoms that make up this universe. These theories have been proved many times in history on the basis of Socrates Atomism Theory, Feynman's Diagram of Quantum Structures, Maxwell's Equation, The Multiverse Theory, General Relativity, and Alfred Morgan's Theory of Continuations in Fluid Mechanics. This belief in Quantum Field Theory also brings the belief on Gluons which are particles that exchange forces between Quarks in the Quantum model. Quarks are the constituents of matter that form hadrons when interacted between the forces of Gluons.

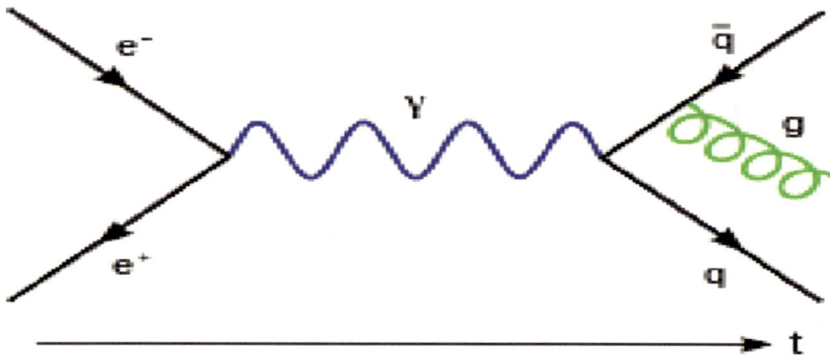

The Dirac Equation is written as:
$$\left(\beta mc^2 + \sum_{k=1}^{3} \alpha_k p_k \, c \right) \psi(\mathbf{x}, t) = i\hbar \frac{\partial \psi(\mathbf{x}, t)}{\partial t}$$

- $\psi = \psi(\mathbf{x}, t)$ is a complex four-component field thought to be the wave function on an electron
- $\mathbf{x}$ and t are the coordinates of space and time,
- m is equal to the rest mass of an electron,
- p is momentum, which is the momentum operator in Schrödinger's theory,
- c is the speed of light, while the formation $\hbar = h/2\pi$ is reduced as Planck constant.

Modern Textbooks write the Dirac Equation as:
$$\left(c\boldsymbol{\alpha} \cdot \mathbf{p} + \beta mc^2 \right) \psi = i\hbar \frac{\partial \psi}{\partial t} .$$

However the way modern textbooks wrote Dirac's equation is an estimate on the real equation itself, not the actual value stated in Dirac's original equation. This makes the modern textbooks' version slightly less accurate.

Canonical Quantization in field theory is known to be the analogous to construction of Classical mechanics as compared to Quantum mechanics as well. The canonical function represents space and time and stays at a specific momentum controlled by gravity. This can lead to the theory of multiple derivatives in the measure of gravity depending on specific location, which matches as being proven by General Relativity. This can also result in a measuring of covariant results in Quantization which is measure of space and time in which it chooses a Hamiltonian structure. A Hamiltonian structure is known as the operator of energy in a Quantum System.

This can be viewed using the formation of Schrödinger's wave mechanics and a view of kinetic energy and momentum in the equation as follows:
$$\hat{T} = \frac{\hat{\mathbf{P}} \cdot \hat{\mathbf{P}}}{2m} = \frac{\hat{p}^2}{2m} = -\frac{\hbar^2}{2m}\nabla^2,$$

$$\hat{H} = \hat{T} + \hat{V}$$
$$= \frac{\hat{\mathbf{P}} \cdot \hat{\mathbf{P}}}{2m} + V(\mathbf{r}, t)$$

which can then be formalized to:
$$= -\frac{\hbar^2}{2m}\nabla^2 + V(\mathbf{r}, t)$$

This is extended to N being the measurement with V being potential Energy and M denoting the mass of particles, and then this equation will be equal to:

$$\hat{H} = \sum_{n=1}^{N} \hat{T}_n + V$$
$$= \sum_{n=1}^{N} \frac{\hat{\mathbf{P}}_n \cdot \hat{\mathbf{P}}_n}{2m_n} + V(\mathbf{r}_1, \mathbf{r}_2 \cdots \mathbf{r}_N, t)$$
$$= -\frac{\hbar^2}{2} \sum_{n=1}^{N} \frac{1}{m_n}\nabla_n^2 + V(\mathbf{r}_1, \mathbf{r}_2 \cdots \mathbf{r}_N, t)$$

This means that for non-interacting particles then the equation is being represented as:

$$V = \sum_{i=1}^{N} V(\mathbf{r}_i, t) = V(\mathbf{r}_1, t) + V(\mathbf{r}_2, t) + \cdots + V(\mathbf{r}_N, t)$$

Schrödinger's measurement of space and time is then viewed as:
$$H \left| \psi(t) \right\rangle = i\hbar \frac{\partial}{\partial t} \left| \psi(t) \right\rangle.$$

which in Dirac's formulation as measure of eigenvectors denoting at a in the spectrum of energy levels being allowed, the equation can then be viewed as:
$$H \left| a \right\rangle = E_a \left| a \right\rangle.$$

As I mentioned earlier, Quarks are the constituents of matter. They can be viewed by various properties including electric charge, color change, mass, spin of energy, and exchange between particles. Quarks are considered elementary particles in the standard model of Quantum Physics which can view them as a different type of operator in condensed matter. Operator meaning source in this occasion; Quarks are also Fermionic meaning their measured by the sub classification of Fermi. Thus the classification of Fermi in particle physics can then be built up into three generations of matter. This will then provide a general understanding in the fundamentals of the field. For example Neutrinos are extremely light sub atomics particles which can often be found in extremely light elements such as helium.

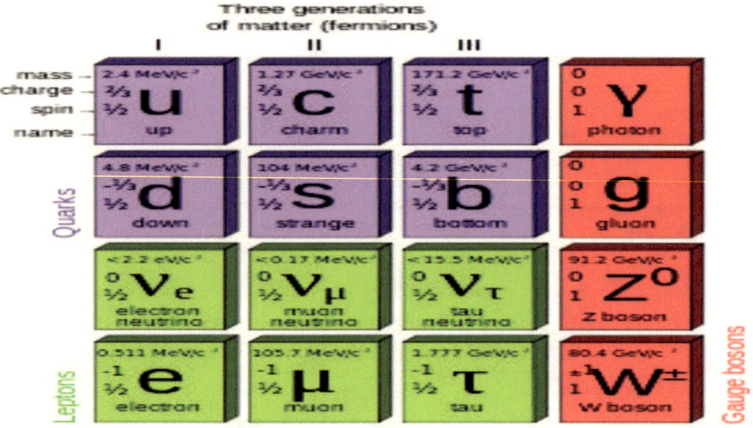

These fundamentals led to the discovery of baryon as seen here:

Next we look at the Quark Flavor Properties seen such as:

Quark flavor properties[67]

Name	Symbol	Mass (MeV/c^2)*	J	B	Q	I_3	C	S	T	B'	Antiparticle	Antiparticle symbol
First generation												
Up	u	1.7 to 3.1	½	+⅓	+⅔	+½	0	0	0	0	Antiup	u
Down	d	4.1 to 5.7	½	+⅓	−⅓	−½	0	0	0	0	Antidown	d
Second generation												
Charm	c	1,290+50 −110	½	+⅓	+⅔	0	+1	0	0	0	Anticharm	c
Strange	s	100+30 −20	½	+⅓	−⅓	0	0	−1	0	0	Antistrange	s
Third generation												
Top	t	172,900±600 ± 900	½	+⅓	+⅔	0	0	0	+1	0	Antitop	t
Bottom	b	4,190+180 −60	½	+⅓	−⅓	0	0	0	0	−1	Antibottom	b

J = total angular momentum, B = baryon number, Q = electric charge, I_3 = isospin, C = charm, S = strangeness, T = topness, B' = botto

We can include that if there is an antibottom then there must be an Antibaryon as well as positive mesons and Baryons as seen on the graphs.

Conclusion: This then can lead to the possibility of neutrinos or toas being extremely light enough that they may have less mass or matter condemned in them then light itself, which can lead to the discovery of their being sub atomic particles faster than the speed of light itself. However if that is true, these equations may steel remain correct.

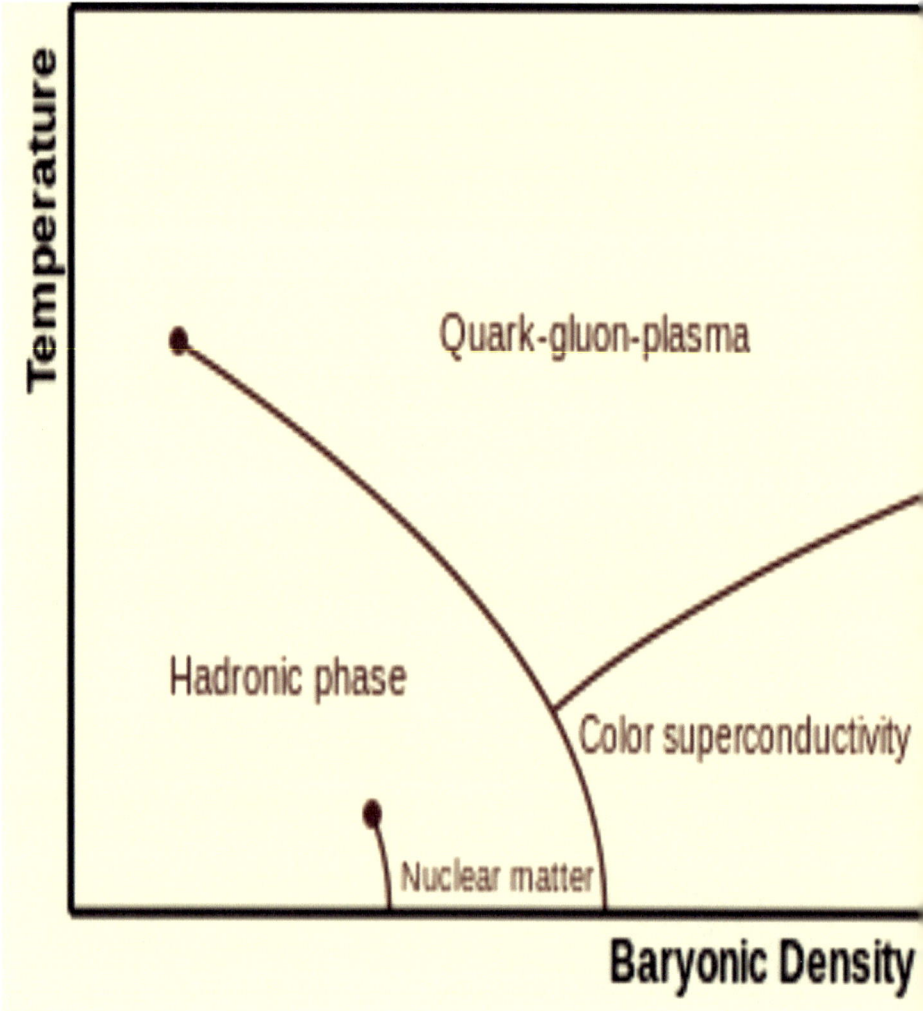

Sources:

Bogoliubov, N.; Shirkov, D. (1982). *Quantum Fields*. Benjamin-Cummings. ISBN 0-8053-0983-7.

^ G. Fraser (2006). *The New Physics for the Twenty-First Century*. Cambridge University Press. p. 91. ISBN 0-521-81600-9.

^ Weiner, Richard M. (2010). "The Mysteries of Fermions". *International Journal of Theoretical Physics* **49** (5): 1174–1180. arXiv:0901.3816. Bibcode 2010IJTP...49.1174W. doi:10.1007/s10773-010-0292-7.

^ "Definition of Faint Hubble Blob". Merriam webster. Retrieved November 6, 2012.

^ Tsetkov, Pavel; Usman, Shoaib (2011). Krivit, Steven. ed. *Nuclear Energy Encyclopedia: Science, Technology, and Applications*. Hoboken, NJ: Wiley. pp. 48; 85. ISBN 978-0-470-89439-2.
.